The Institute of Biology's
Studies in Biology no. 44

Biology of Aphids

A. F. G. Dixon

B.Sc., D.Phil.
Senior Lecturer in Zoology,
University of Glasgow

Edward Arnold

© A. F. G. Dixon 1973

First published 1973
by Edward Arnold (Publishers) Limited,
25 Hill Street,
London W1X 8LL

Board edition ISBN: 0 7131 2421 0
Paper edition ISBN: 0 7131 2422 9

To
June Dixon
and
Dick Hille Ris Lambers

Printed in Great Britain by
The Camelot Press Ltd, London and Southampton

General Preface to the Series

It is no longer possible for one textbook to cover the whole field of Biology and to remain sufficiently up-to-date. At the same time teachers and students at school, college or university need to keep abreast of recent trends and know where the most significant developments are taking place.

To meet the need for this progressive approach the Institute of Biology has for some years sponsored this series of booklets dealing with subjects specially selected by a panel of editors. The enthusiastic acceptance of the series by teachers and students at school, college and university shows the usefulness of the books in providing a clear and up-to-date coverage of topics, particularly in areas of research and changing views.

Among features of the series are the attention given to methods, the inclusion of a selected list of books for further reading and, wherever possible, suggestions for practical work.

Readers' comments will be welcomed by the author or the Education Officer of the Institute.

1973

The Institute of Biology,
41 Queens Gate,
London,
SW7 5HU

Preface

Aphids are serious pests of agricultural crops and forest trees. The study of a pest species of aphid undertaken as part of the International Biological Programme indicates the importance of aphids on a world-wide scale. Aphids as plant feeders are an important component of most terrestrial communities in temperate regions and many other organisms are dependent upon them for food.

To give a short account of aphid biology it has been necessary to generalize and thus obscure to a great extent the adaptability and variability shown by the different species. Nevertheless, it is hoped that this account will serve to excite interest in aphids and their biology.

I would like to thank Drs. T. J. Dixon and J. M. S. Forrest, and Mr. J. Shearer for reading the manuscript and making helpful suggestions. I am also indebted to several authors and editors for permission to use illustrations from their papers. I am especially grateful to Professor M. H. Zimmermann for the excellent photographs for Plates 1 and 2, and to Dr. J. A. Dunn for the photographs for Plate 3. Mr. N. Kidd gave much appreciated help by drawing Figures 1-2, 5-1 and 7-4.

1973   A. F. G. D.

Contents

1 The Salient Features of Aphids

1.1 Introduction

Aphids are an extremely successful group which occurs throughout the world, with the greatest number of species in the temperate regions. As individuals they are small and inconspicuous. However, they frequently become so numerous that the number feeding on the leaves and shoots on 0·4 hectares (an acre) of ground is 2000 million. The roots of plants may support a further 260 million. Many species are agricultural pests, and tree dwelling aphids can severely retard their host plants. Several generations are born a year and their fecundity is great. Their complex life cycles and polymorphism enable them to exploit their host plants and respond to every contingency of their environment to a high degree. They can migrate great distances, up to 1300 km, and in the temperate regions few plant species are without a specific aphid.

Aphids originated from the Archescytinidae in the Carboniferous era, or early Permian, 280 million years ago (HEIE, 1967). A conspicuous evolution of aphids was later associated with the appearance of flowering plants, the Angiosperms, and the host plants of most present-day aphids are the Angiosperms, although some aphids live on Gymnosperms and a few species attack ferns and mosses.

Aphids belong to the insect superfamily Aphidoidea, within the order Homoptera, the plant-sucking bugs. The Aphidoidea are all soft bodied insects, whose wings if present are membraneous; these insects all feed on plant sap. Aphids are distinguished from other groups in the Aphidoidea in that the females, of at least a few generations, do not require fertilization for the development of their embryos: they are parthenogenetic; and also in that these asexual females produce live offspring: they are viviparous. The asexual females can possess wings like most adult insects, or lack them. Winged aphids are known as alatae and wingless aphids as apterae. Typically there are many structurally different morphs in a species, including both sexual and asexual forms. Polymorphism, which is the development within a species of several different adult morphs, is characteristic of aphids.

1.2 Mouthparts: their structure and mode of operation

Aphids feed by inserting their slender mouthparts, the stylets, into plant cells. The stylet bundle consists of an outer pair of mandibular

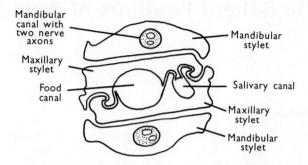

Fig. 1-1   Diagram of a transverse section through the stylet bundle of an aphid

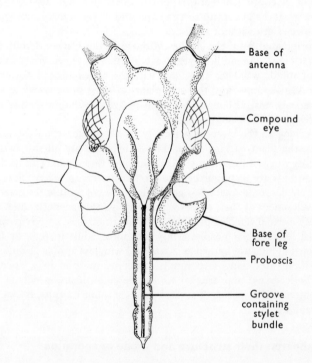

Fig. 1-2   Anterior view of the head and proboscis of an aphid to show the groove in the proboscis in which lies the stylet bundle

stylets and an inner pair of maxillary stylets (Fig. 1–1). The maxillary stylets are held together by interlocking grooves and ridges, and in channels between them run the food and salivary canals. In the resting position the stylet bundle lies in a groove along the anterior surface of the aphid's proboscis. The terminal segment of the proboscis grips the stylet bundle (Fig. 1–2). When the stylets are inserted into the tissues of a plant the second segment of the proboscis slides back along the stylets and telescopes within the wider and softer basal first segment. In aphids which have a very long proboscis this invagination can result in the first and second segments of the proboscis lying within the aphid's abdomen. When penetrating the tissues of a plant the stylets go between the cells, and only rarely pass through a cell, until they reach the sieve tubes within the veins of the host plant. The stylets then pierce the cells and feeding commences (Plate 1). Penetration by the stylets is achieved by alternate protraction of first, the mandibular, and then the maxillary stylets. Each mandibular stylet tapers to about 0·04 μm or less in diameter at the tip. Because of this shape any pressure applied to the relatively large base of the stylets is transformed into a much greater pressure per unit area at the tip. Thus mandibular stylets are well adapted to pierce and penetrate between plant cells. Penetration is assisted further by secretion of saliva which passes down the salivary duct in the maxillary stylets and is exuded from the tip of the stylet bundle. A pectinase in the saliva breaks down the bonding between the cells.

While an aphid is probing within plant tissues it secretes a stylet sheath around the stylets (Fig. 1–3). The mainly protein sheath material is secreted by the salivary glands (Fig. 1–4). Probably the most important function of

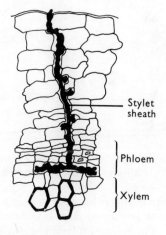

Fig. 1–3   Stylet sheath laid down in the tissues of a plant by an aphid feeding on the phloem sieve tubes

the sheath is to give rigidity to the very flexible stylet bundle and thus restrict bending to the apex of the stylets (POLLARD, 1969).

When moving between cells the stylets often change direction. The maxillary stylets are locked together, and so changes in direction are achieved by the unequal contraction of the muscles attached to the base of each maxillary stylet. By rotating on its proboscis, and thus rotating the stylets, an aphid can move the tips of its stylets through 360° in a plane at right angles to the line of penetration of the stylet bundle. In this way it can explore an extensive area of plant tissue without withdrawing its stylets.

1.3   Ingestion and digestion of plant sap

Most aphids obtain their food from the sieve tubes in the phloem tissue of plants. The sap in these cells is under considerable pressure and probably aphids rely on this turgor pressure to force sap up the very fine food canal in the stylets. However, turgor pressure does not account for aphids feeding successfully on artificial diets which are not under pressure. In this case aphids probably use the cibarial pump to suck up their food (Fig. 1–4).

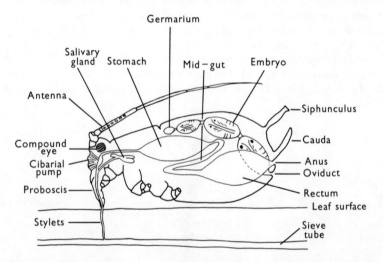

Fig. 1–4   Diagram of an aphid feeding on a plant to illustrate the internal anatomy relative to the external features of the aphid

Phloem sap is rich in sugars but poor in amino-acids, which are essential for growth. Aphids ingest a very large amount of food in order to acquire sufficient protein. The residual solution of digested food, mainly sugars, is stored in the dilated rectum before ejection to the exterior in the form of a droplet of honeydew (Plate 2). Some aphids are attended by ants which collect and feed on honeydew. When aphids are abundant the leaves of their host plant become coated with sticky honeydew. In some of the drier parts of the world large quantities crystallize into manna and this is collected and eaten by man.

1.4   Nitrogenous metabolism and excretion

The excretory system in most insects consists of Malpighian tubules, which regulate the balance of water and ions in the body fluids, and remove nitrogenous waste. However, aphids lack Malpighian tubules, and excrete nitrogenous waste in the form of ammonia instead of uric acid (LAMB, 1959). The large volume of water in the diet is apparently sufficient to dilute and flush out ammonia through the gut and obviate the toxic effects. Symbiotic organisms within certain cells of the body are thought to detoxify part of the ammonia produced by the aphid and also to fix atmospheric nitrogen. The cells containing these organisms are called mycetomes, and are extremely well developed in very young aphids.

It has been suggested that aphids compensate for the poor protein content of their diet by digesting their symbionts. TÓTH (1940) based this suggestion on the observation that mycetomes increase in number and symbionts are dissolved in aphids that are growing rapidly. However, a detailed study of the nitrogen economy of the willow aphid has revealed that in this species all the nitrogen it needs can be obtained from the food it ingests (MITTLER, 1958). Thus the role of the symbionts in the life of an aphid remains unknown.

1.5   Gonads and embryogenesis

One of the remarkable features of aphids is their mode of reproduction. They are parthenogenetic and viviparous (Plate 3) for most of the year, but are also capable of sexual reproduction with the production of eggs. The gonads of a female aphid consist of two ovaries each composed of from 4 to 6 ovarioles. At the distal end of each ovariole is a germarium from which the eggs are ovulated (Fig. 1–4). As soon as they are ovulated eggs begin development in the parthenogenetic female without fertilization, whereas those in sexual females require fertilization. Even ovarioles of newly born parthenogenetic females contain embryos. Therefore, a mother can have in its ovarioles, developing embryos which in turn also contain embryos, the future granddaughters. Telescoping of generations is made possible by parthenogenesis and viviparity and is common in aphids. The embryos or eggs pass to the exterior via an oviduct.

All aphids appear to show diploid parthenogenesis. That is, there is no reduction division and development starts from a cell containing a full complement of chromosomes including 2X chromosomes. Males are produced by the loss of an X chromosome at the first division. Males produce haploid sperm which contain an X chromosome only. Fertilization of the haploid eggs of the sexual females results in a diploid cell with two X chromosomes, and therefore all aphids hatching from fertilized eggs are females.

The offspring of viviparous females are born rear first and fully active, and when deposited, face in the direction from which predators are most likely to approach. At birth they are structurally very similar to an adult. The nymphs destined to become winged adults possess wing buds. At each of the four moults there is an increase in size and a gradual change towards the adult form. There is no dramatic transition in form between the last juvenile stage and the adult, as in many other insects. In appearance the apterous morph is more like a juvenile than is an alate. In this neotenous state apterae can exploit a favourable situation. However, with the onset of adverse conditions aphids develop wings and seek out more favourable areas. Parthenogenesis enables those aphids that survive to produce others like themselves, also capable of surviving in that environment, and thus rapidly to increase in numbers.

2 Life Cycles and Polymorphism

2.1 Life cycles

The production of both sexual morphs as well as parthenogenetic virgino-parae is common in aphids in temperate regions. Sexual morphs are produced mainly in the autumn and, after mating, the oviparous females lay the overwintering eggs. The following year, when plants resume growth, eggs hatch and a series of parthenogenetic generations develop.

Most aphids live on only one species of host plant. The sycamore aphid belongs to this group and feeds on sycamore trees, *Acer pseudoplatanus* L. It spends the winter months in the egg stage. In spring when the buds of sycamore begin to develop, the eggs hatch and the nymphs develop into the winged adults of the first generation, known as fundatrices. These adults are parthenogenetic and give rise to nymphs which develop into parthenogenetic virginoparae. Several parthenogenetic generations occur in succession until the onset of autumn when the nymphs then develop into apterous egg-laying females and alate males. The sexual morphs mate and the oviparae lay the overwintering eggs (Fig. 2–1). When a tree be-comes heavily infested, sycamore aphids will fly from their host plant to colonize other sycamore trees. However, the aphids colonize only other sycamore trees: the aphid is monophagous.

In contrast to monophagous aphids there are polyphagous aphids which live on more than one species of host plant. The bird cherry-oat aphid spends autumn, winter and spring on bird cherry trees, *Prunus padus* L., but in summer migrates to various alternative grass species. The funda-trices hatch from the overwintering eggs as the buds of bird cherry burst in spring. The offspring of the fundatrices mainly develop into apterous adults, and more rarely into alatae. The aphids in this generation are referred to as fundatrigeniae and are distinct from the fundatrices. The alate fundatrigeniae and all of the third generation, also composed of alates, comprise the emigrants which fly away from bird cherry and colonize various grasses. The aphid then passes through several generations on grass. When conditions become overcrowded alate individuals develop and fly off and colonize other grass plants. The generations restricted to grass are referred to as exules. In autumn alate gynoparae and alate males develop on grass. These fly to bird cherry where the gynoparae give birth to ovi-parae which complete their development on bird cherry. The males mate with the oviparae, which then lay their eggs around the buds and in crevices in the bark of bird cherry. This seasonal movement between a woody

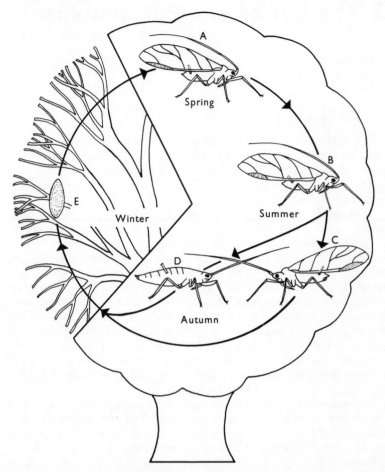

Fig. 2–1 Life cycle of the sycamore aphid (A, fundatrix; B, alate virgino-
para; C, male; D, ovipara and E, egg)

primary host and an herbaceous secondary host is known as host alternation. Aphids that show this type of life cycle have many different adult
morphs; i.e. they are highly polymorphic (Fig. 2–2).

From host alternating or dioecious aphids have evolved species that
complete their life cycle on what was originally a secondary host plant.
Such aphids are often also polyphagous, feeding on a number of different
species of host, as do the exules of some of the host alternating aphids.
Some of these aphids do not produce sexuals but exist parthenogenetically
all through the year. Such aphids which have no egg stage are referred
to as anholocyclic. In several species of host alternating aphids there exist

both holocyclic and anholocyclic clones. Mild winters allow survival of parthenogenetic forms but they are killed in severe winters. After severe winters anholocyclic clones develop again from the holocyclic clones (BLACKMAN, 1971). Aphids which damage agricultural crops are mainly host alternating, or polyphagous non-host alternating species, and this has obscured the fact that most aphids are monophagous.

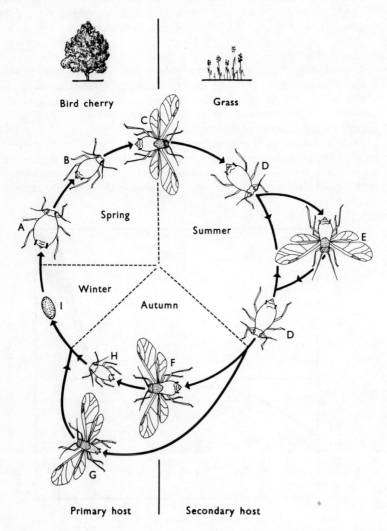

Fig. 2–2 Life cycle of the bird cherry-oat aphid (A, fundatrix; B, apterous fundatrigenia; C, emigrant; D, apterous exule; E, alate exule; F, gynopara, G, male; H, ovipara and I, egg)

2.2   Factors affecting morph determination

For some species of aphid it has been suggested that the seasonal sequence of morphs is predetermined and linked to a timing mechanism. However, all the species of aphid studied indicate that the seasonal development of the different morphs is initiated primarily by changes in the external environment, and internal changes in aphids govern the responsiveness of the aphid to changes in its environment.

2.2.1   Sexual morphs

Towards autumn the days become shorter and cooler. In 1924 MARCO-VITCH showed that oviparae of the strawberry root aphid could be induced to develop in summer by shortening normal day-length, and conversely gamic reproduction could be postponed by extending the normal daylight period in the autumn. This is the first record of photoperiodism in animals. It is now known that of the factors in the environment of an aphid, day-length, temperature and the quality of the host are important in the induction of sexual morphs. In addition internal changes in the aphid also influence the aphid's response to changes in its environment.

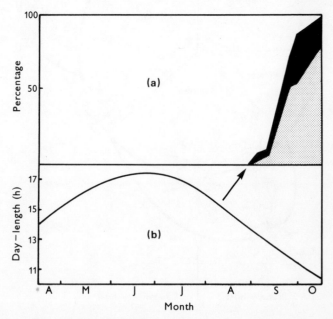

Fig. 2-3 The relationship between **(a)** the appearance in the field of the different morphs in the sycamore aphid and **(b)** day-length (☐ , virginoparae; ▒ , ovidparae; ■ , males)

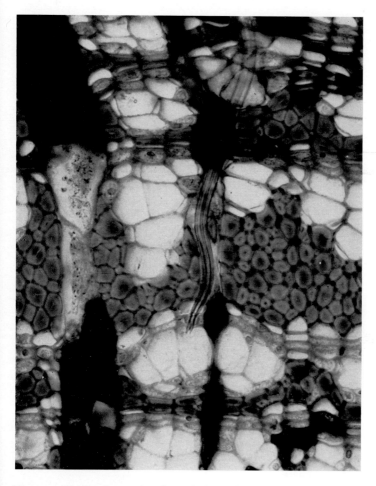

Plate 1 Transverse section through the bark of lime (*Tilia americana* L.) showing the tips of the stylet bundle of *Longistigma caryae* Harris which have penetrated a sieve element. (Photographed by Professor M. H. Zimmermann)

Plate 2 *Longistigma caryae* Harris excreting a droplet of honeydew. (Photographed by Professor M. H. Zimmermann)

In northern parts of the United Kingdom the sexual morphs of the sycamore aphid are born in the first week of August when day-length is 15·5 h and they reach maturity in early September (Fig. 2–3). Short day-length induces the aphid to give birth to oviparae. No oviparae appear if the aphid is kept under long day conditions. Males do not appear until the fourth generation, and their appearance does not appear to be related to day-length. The appearance of males only after a certain number of generations has elapsed, and apparently uninfluenced by extrinsic factors, appears to be characteristic of monoecious species of aphid. Although controlled in different ways the maturation of males and sexual females of the sycamore aphid is synchronized.

In highly polymorphic host alternating species of aphid, such as the bird cherry-oat aphid, the appearance of both of the sexual morphs is induced by short day-length (Fig. 2–4). Under short day conditions the

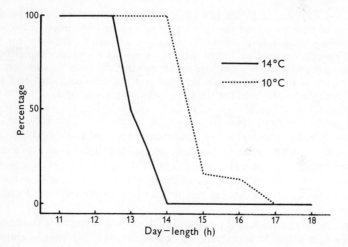

Fig. 2–4 The relationship between the percentage of sexual morphs produced by the bird cherry-oat aphid and day-length, at 10°C and 14°C

bird cherry-oat aphid gives birth first to gynoparae which give rise to oviparae, and then in the second half of the mother's reproductive life to males. Oviparae are born to the first gynoparae by the time males are being born.

The photoperiodic receptor concerned with morph determination is located in the mid-dorsal region of the head (Fig. 2–5) and does not involve the compound eyes of the aphid (LEES, 1964). As there are no specialized receptor organs in the cuticle of this part of the head, it is thought that the photoreceptor is located in the brain of the aphid. Light penetrates the cuticle and stimulates the photoreceptor directly.

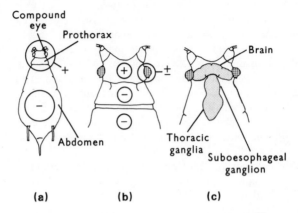

Fig. 2–5 Region of an aphid's body sensitive to day-length. Using fine beams of light to extend the day-length experienced by different parts of an aphid's body only prevents the induction of sexual morphs when the head plus thorax are illuminated (**a**). Increasing the day-length experienced by small areas of the head and thorax reveals that the most sensitive area is the region of the head between the eyes and not the compound eyes (**b**). In this region of the head lies the brain (**c**). (After Lees, 1964)

In some species of aphid low temperatures promote the production of sexual morphs. The higher the temperature the bird cherry-oat aphid is exposed to the shorter the day-length necessary to induce the appearance of gynoparae and males (Fig. 2–4). The effect of low temperature serves to increase the production of sexual morphs in autumn. In mild autumns sexual morphs appear later as shorter day-lengths are required for their induction.

In several species sexual production takes place in response to changes in the host plant. In temperate regions short day conditions in autumn induce plants to cease growth and become dormant. Aphids living on the roots of these plants in continuous darkness and at relatively constant temperatures, often at considerable depths, produce sexual morphs as the plant becomes dormant. Aphids that live above ground but produce sexual morphs when day-length is at its greatest respond to the cessation of shoot growth of their host plant (FORREST, 1970). It is unknown what changes, associated with the cessation of growth of the host plant, trigger the changes in the aphid. Whatever the mechanism, it does, like short day conditions and low temperature, enable some aphids to synchronize their life cycle with the growth and development of their host plant.

Although many aphids such as the sycamore aphid hatch from eggs early in the year and are present in the spring when day-length is as short as 14 h oviparae never appear in the field before autumn (Fig. 2–3). The appearance of the sexual morphs is controlled by an intrinsic timing mechanism which in the case of ovipara production controls the sensitivity of the aphid to its environment. Aphids which hatch from the overwintering eggs cannot readily be induced to give rise to sexual morphs, but each successive generation can be more readily induced to give rise to sexuals (Fig. 2–6).

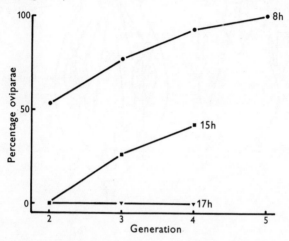

Fig. 2–6   The effect of day-length on the percentage of the female offspring in the sycamore aphid that are oviparae in the second, third, fourth, and fifth generations (●—●, 8-h; ■—■, 15-h; ▼—▼, 17-h day)

The nature of the internal timing mechanism is still a matter of conjecture. LEES (1963) favours a photoperiodic receptor linked with an endocrine organ. Cauterization of certain neurosecretory cells in the brain of two species of aphid has shown that ovipara production is under endocrine control (DEHN, 1969). It is likely that the same neurosecretory cells act as receptor and effector (Fig. 2–7). However, it is not known what controls the receptive state of the receptor.

In conclusion, aphids are sensitive, to varying degrees, to changes in their host plant and their physical environment. Changes within an aphid control its sensitivity to changes in its external environment. In this way aphids synchronize their life cycles with the seasonal development of their host plants.

2.2.2   Parthenogenetic morphs

In addition to the sexual morphs in highly polymorphic species of aphid there are also alate virginoparae which possess wings and can fly, and apterous virginoparae which lack wings. The following factors have been implicated in the development of alate forms, (a) crowding, (b) quality of the host plant, (c) ant attendance, (d) temperature and photoperiod and (e) intrinsic factors.

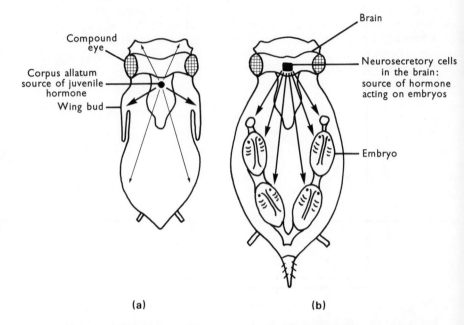

Fig. 2–7   Diagrammatic representation of (**a**) the source of juvenile hormone which affects the development of wings and pigmentation during nymphal development and (**b**), the source of the hormone which acts on embryos before they are born and induces them to develop after birth into sexual forms

Alatae often begin to appear in aphid colonies when the aphid is very numerous (Fig. 2–8) and is having an adverse effect on its host plant. In such situations the development of alatae among the progeny can be induced by crowding or changes in the host plant.

LEES (1967) has shown that crowding is the most important factor inducing the development of alatae in the vetch aphid. Aphids reared in isolation never give rise to alatae even when reared on plants of poor quality. Tactile stimulation associated with crowding induces the development of alatae. In the vetch aphid it is the mother that is responsive to crowding. Rearing in isolation offspring of mothers that have been subjected to crowding does not prevent the offspring from becoming alate. In other species the aphid is responsive to crowding in its postnatal life and in still others both the experience of the mother and offspring are important in determining whether the offspring will become alate.

The period during which the aphid is sensitive to crowding varies from 18-h from the time of birth for the alfalfa aphid (TOBA *et al.*, 1967) to the continuous monitoring of crowding in such aphids as the bird cherry-oat

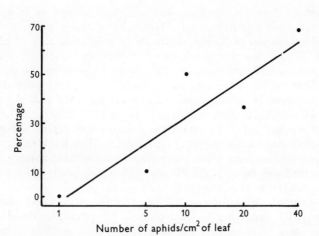

Fig. 2-8 The effect of crowding on the percentage of bird cherry-oat aphids which develop into alatae

aphid (DIXON & GLEN, 1971). In aphids that continuously monitor crowding a period of isolation following a period of crowding can partly reverse the effect of crowding and intermediates develop. These are intermediate in structure between apterae and alatae.

Although the production of alatae is associated with high numbers of the aphid it is not the number of aphids which is important but the degree of tactile stimulation between the aphids. Even two aphids can constitute a crowd if they come into contact with one another.

In some species the response to crowding is influenced by the quality of the host plant. The pea aphid is more likely to respond to crowding if it is feeding on a mature plant (SUTHERLAND, 1969). In other aphids the quality of the host plant has a direct effect on alary dimorphism. With the cessation of shoot growth of bird cherry the bird cherry-oat aphid produces alate emigrants even when reared in isolation. However, crowding enhances the production of alatae (DIXON & GLEN, 1971).

Several species of aphids are tended by ants. However, the tactile stimulation inflicted by ants does not induce the development of winged forms but delays their appearance. Likewise high temperatures and long days may also inhibit the development of alatae. Alatae of the vetch aphid that fly cannot readily be induced to give rise to alatae. Those that do not fly respond to crowding and produce alate offspring as do apterae. This enables alatae that remain in a colony to respond to over-crowding and produce alatae which can fly away.

The development of alate and apterous morphs is under the control of an endocrine gland, the corpus allatum. This gland is situated behind the brain of an aphid and produces juvenile hormone. In nymphs that develop into apterous adults the corpus allatum is more active than in nymphs that become alatae (WHITE, 1971). Treatment of alatoid nymphs by the external application of juvenile hormone or substances that mimic the action of juvenile hormone results in the development of alatae that retain some nymphal characteristics. Juvenile hormone prevents the development and expansion of the wing buds present at birth in many aphids, and the development of melanic pigment in the cuticle of the abdomen (LEES, 1966). Therefore, the amount of juvenile hormone circulating in the body fluids of a parent aphid and its offspring during their postnatal development will determine the morph of the offspring. High levels result in the development of apterous adults which retain many juvenile characteristics, whereas low levels result in the development of alate adults which possess all the adult features (Fig. 2–7).

In conclusion, aphids are sensitive to changes in their biological environment. By responding to these changes and producing alatae which colonize other plants aphids can to a large extent avoid overcrowded conditions and poor nutrition.

2.3   Adaptive significance of apterousness

The winged condition is thought to be the more primitive condition in aphids and the apterous condition to have been a later evolutionary development. The initial advantage of apterousness was an increase in fecundity because the development and maintenance of wing musculature possibly competes with the development of embryos for the limited amount of nitrogen available to the aphid. This is the case in *Drepanosiphum dixoni* H.R.L., where brachyptery occurs. Brachypterous and macropterous alatae in *D. dixoni* are similar except that the brachypterous alatae cannot fly as they lack indirect wing muscles, and their wings, although perfect, are shorter than those of macropterous alatae. Brachypterous alatae are 32% more fecund than macropterous alatae (DIXON, 1972).

In the more highly polymorphic species the morphology of the apterae and alatae is very different. The alatae being colonizers have been subject to different selection pressures from those acting on apterae. This may account for alatae producing smaller offspring and also a larger proportion of their offspring early in adult life than do apterae. By producing many small offspring and by giving birth to most of their offspring soon after maturation alatae can maintain a rapid rate of increase, which in certain species is as great as that of the apterae. Because of their wings alatae can more easily disperse to exploit fresh host plants, and, associated with the winged condition, they are also capable of quickly producing the next generation of mature adults. However, the first change in the evolution of apterousness was possibly a simple increase in fecundity of brachypterous forms as seen in *D. dixoni*.

3 Host Alternation

3.1 Introduction

In temperate regions host alternation is common amongst aphids and is associated with seasonal changes in the weather and growth of plants, in these regions. Aphids are weak fliers and are likely to be carried on the wind rather than to have control of the direction of their flight. To leave a host plant and seek out not only another host plant, but another species of host plant which may be sparsely distributed, is hazardous. Nevertheless, host alternation is a successful mode of life and therefore there are advantages to such behaviour.

3.2 Adaptive significance of host alternation

Among the aphids that alternate between a winter woody host plant and summer herbaceous hosts is the bird cherry-oat aphid. Emigrants of the bird cherry-oat aphid leave bird cherry before the amount of soluble nitrogen in the leaves falls below 0·4%, and the gynoparae and males return in September when the level has risen again. In fact, the period of absence of this aphid from bird cherry coincides with the period when the leaves are mature and provide a poor source of food for the aphid (Fig. 3–1).

Both DAVIDSON (1927) and MORDWILKO (1928) have stressed that seasonal movement in host-alternating aphids enables them to exploit a continuous supply of nutritionally favourable foliage which is either growing or senescent. When shrubs and trees are either actively growing, or the leaves are approaching senescence, the phloem sap contains relatively high concentrations of amino-nitrogen in the form of many amino-acids. In summer when growth has ceased the sap is poor in amino-nitrogen and contains few amino-acids. In the sycamore aphid, which lives all year on sycamore, both the aphid's size and reproductive activity are determined by the quality of their food. As the concentration of amino-nitrogen in the phloem sap falls in spring, the nymphs develop into progressively smaller adults, the birth rate drops, and reproduction may cease altogether while the leaves are mature. When the leaves become senescent in autumn, reproduction recommences, the growing aphids reach a larger size, and the reproductive rate increases (Fig. 3–2). In summer when the leaves and shoots of many shrubs and trees have ceased growing and provide poor food for aphids, herbaceous plants continue to grow and therefore offer a more nutritious source of food to the polyphagous species.

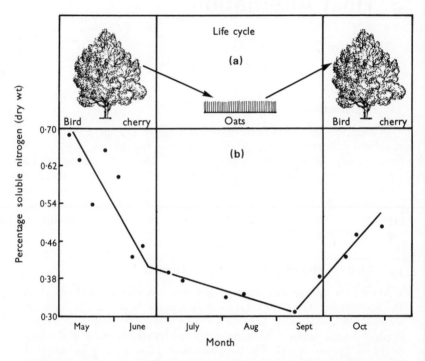

Fig. 3–1   The relationship between (**a**) the movement of the bird cherry-oat aphid from bird cherry to oats in early summer and from oats to bird cherry in autumn and (**b**) the percentage soluble nitrogen present in the leaves of bird cherry

Moreover, in spring there is a very rapid increase in the number of insect predators on woody plants, feeding on aggregated host alternating aphids like the bird cherry-oat aphid. This coincides with the approaching end of the aphid's stay on the primary host (Fig. 3–3). In spite of the assemblage of predators the number of bird cherry-oat aphids diminishes at this time, largely because of the departure of alate emigrant aphids produced in response to crowding and change in nutrition (Fig. 3–3c), although some mortality is caused by predators. Predators which remain after the aphids have flown away resort to cannibalism and few survive. Therefore, because they move to another host, aphids effectively evade their natural enemies.

Aphids that show host alternation characteristically aggregate in dense colonies on the growing points of their host plants, and are therefore more conspicuous and vulnerable to their enemies. When a large aphid population has developed on a plant, some of the alates escape from the large accumulation of predators and parasites by moving to another host plant.

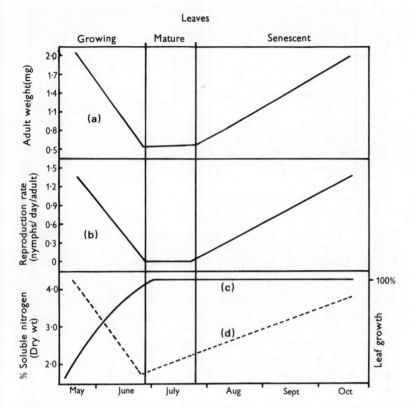

Fig. 3–2 The relationship between the progress of growth and maturation of sycamore leaves and the weight (**a**) and reproductive rate (**b**) of the sycamore aphid (**c** = leaf growth, **d** = percentage soluble nitrogen present in the leaves)

These aphids escape successfully because they move to another plant to which enemies have not yet been attracted. They establish a new colony, which, because of its isolation from other colonies, is unlikely to be located quickly by enemies irrespective of their numbers. While the aphid remains undetected it can multiply quickly, because of the rich food supply, producing more alates to colonize other plants. However, once located, dense aggregations of aphids like the black bean aphid are easy prey and are quickly destroyed by predators. Movement from plant to plant would certainly enable an aphid to exploit the spatial heterogeneity that exists in nature, and so avoid annihilation by predators and parasites (DIXON, 1971a).

Many aphids are monophagous and tree dwelling, and some of these species can reproduce even when the leaves of their host are mature.

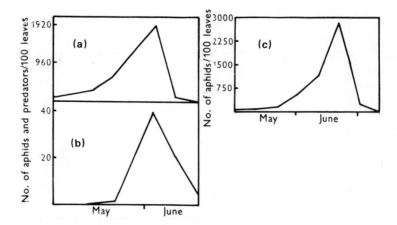

Fig. 3-3 The number of (**a**) bird cherry-oat aphids and (**b**) insect predators present in May and June on bird cherry in the field, and (**c**) the number of bird cherry-oat aphids in May and June in the absence of predators and parasites

Although they may aggregate on certain leaves, characteristically they space out. Because neither their visual field nor their path of escape is obstructed by other aphids and because of the alacrity of these aphids, an approaching predator can be seen and avoided (p. 46).

Lime, *Tilia sp.*, is attacked by two species of aphid, both feeding on the leaves. *Patchiella réaumuri* (Kalt.) forms dense aggregations inside leaf

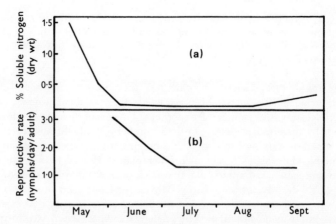

Fig. 3-4 The relationship between (**a**) the percentage soluble nitrogen present in lime leaves and (**b**) the reproductive rate of the lime aphid

pseudogalls in spring. When the lime leaves mature it colonizes *Arum maculatum* L. and returns to lime in autumn when lime leaves approach senescence. The other aphid, *Eucallipterus tiliae* (L.) remains on lime all year. In this species the aphids are spaced apart and leap off the leaves when disturbed. Insect predators find them difficult to catch. In addition the aphid continues to reproduce even when it feeds on mature leaves (Fig. 3–4). Hence faced with poor nutritional conditions when the host plant matures, and also with increased numbers of enemies, relatively inactive aphids that are committed to live in dense colonies may find host alternation a means of evading natural enemies.

Quite exceptional are the host alternating aphids of the family Fordinae. The Fordinae exploit the woody primary host in summer and colonize grasses during the winter, the reverse of the usual behaviour in aphids. This is the way in which Fordinae thrive in areas with arid summers and moist winters.

MORDWILKO (1928) regarded host alternation as an evolutionary end point for aphids. However, aphids have evolved a great variety of life cycles, and host alternation is but one of many strategies that have been evolved to exploit plants.

4 Migration

4.1 Introduction

Since aphids move from plant to plant most conspicuously by flight (Fig. 4–1) a great deal of research has been done on this. The early literature treated aphid species as migratory only if they showed host alternation. KENNEDY'S (1961*a*) definition of migration as a 'persistent, straightened-out

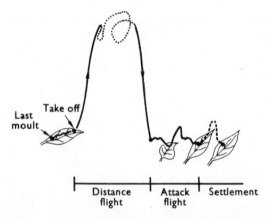

Fig. 4–1 Diagrammatic representation of take-off, flight and settling in aphids. (After MOERICKE, 1955)

movement that is accompanied by and dependent upon the maintenance of an internal inhibition of those vegetative reflexes that will eventually arrest movement' stresses the common features of migratory behaviour in animals as different as aphids and birds, and is the one followed here.

4.2 Factors that determine flight

Until recently it was assumed that all alates of highly polymorphic species of aphid were obligate migrants. However, the degree of over-crowding at the time of the adult moult affects how many vetch aphids will fly, and the extent of the overcrowding during both the parental and nymphal life of the bean aphid will influence how many of this species will fly before reproduction takes place (DIXON *et al.*, 1968; SHAW, 1970). Flight in the lime aphid also depends upon previous and current experience.

Flight occurs before the onset of reproduction even in the absence of other aphids if conditions of overcrowding in the past have been severe. Crowding after the adult moult is essential for flight to take place, if the aphid has no previous experience of crowding. In the sycamore aphid previous experience of overcrowding has no effect on the flight behaviour of the adults. They respond directly to current conditions of crowding and nutrition (DIXON, 1969). That flight activity of the bean and vetch aphids is influenced by earlier conditions is, like alate production, of adaptive significance, as high numbers of these aphids have a very marked adverse effect on their host plants. Sycamore aphids do not affect their host plant to the same extent, and thus past population conditions are not as important.

JOHNSON (1969) and his colleagues demonstrated the factors which determine when flight takes place and how many aphids fly. It was previously thought that weather, in particular high wind speeds and low humidity, inhibited the take-off behaviour of aphids and were the prime cause of changes in the numbers of aphids flying.

During the day the air just above the ground is warmest and tends to rise. Therefore, aphids flying during the day are likely to be carried into the upper air by convection currents. Consequently most flying aphids are at heights in excess of 30 m. There is a great similarity between the mechanics of aphid transport by turbulent air motion and the turbulent diffusion of inert particles. Aphids are likely to be carried to high altitudes and subsequently brought back to ground level almost entirely by atmospheric circulation. At sunset, the air just above the ground becomes colder than the air above. Convection then ceases and atmospheric circulation gradually subsides. In southern England, this results in the aphids coming down to ground level, and few or none remain airborne overnight. Atmospheric circulation dominates the aphids' movements because of their low flight speed, 1·6 to 3·2 km h^{-1}, relative to the speed of movement of the air in which they are flying. As the air in which they fly usually moves faster than 3 km h^{-1} the aphids have scarcely any control over where they are going, and can be carried for great distances. They can settle only if the air movement brings them down to ground level, or if they stop flying.

As few aphids remain airborne over southern England at night the aerial population is built up anew each day. Usually the number of aphids in the air even at a height over 600 m has a bimodal periodicity, with a peak in numbers in mid-morning and another in the early afternoon (Fig. 4–2a). JOHNSON et al. (1957) have explained the black bean aphid's flight periodicity. The number flying each day is governed by the number moulting to the adult stage, the length of the teneral period of the adult alate aphid, and the weather affecting take-off. Moulting to the adult stage shows a characteristic daily periodicity. There is a high rate of moulting in the early morning, followed by a lower rate during the rest

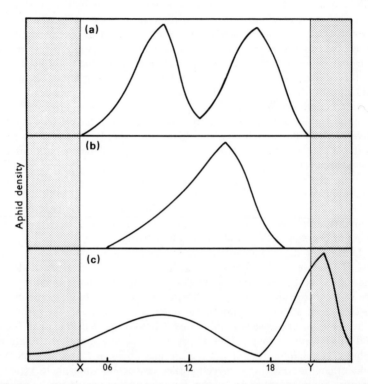

Fig. 4–2 Flight curves of (a) the black bean aphid, (b) the lime aphid and (c) the sycamore aphid. X and Y, time of sunrise and sunset, respectively. ((a) and (b) after LEWIS & TAYLOR, 1965)

of the day, sometimes rising again in the evening and falling almost to zero during the night (Fig. 4–3). This moulting rhythm is thought to be controlled by exogenous factors, in particular temperature. The teneral period between final ecdysis and first flight of the adult alate is temperature dependent, with aphids flying much sooner at higher temperatures.

Aphids become flight mature all through the day and night, temperature permitting. When light intensity and temperature rise above the threshold for take-off, aphids that mature overnight, but are prevented from flying by cold and darkness, take off (Fig. 4–4) and give rise to the first flight peak of the day. The second flight peak later in the day represents the peak in number of newly moulted adult aphids that appear in the morning and complete their teneral development by afternoon. The second flight peak declines in the evening because as the temperature drops fewer nymphs moult to become adult, and also because light fades to intensities below the threshold necessary for take-off. Most of the aphids in the air

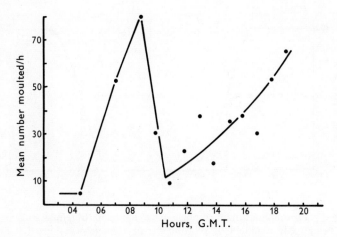

Fig. 4–3   Mean number of fourth instar nymphs of the black bean aphid which moult into alate adults during a day. (After JOHNSON, HAINE, COCKBAIN & TAYLOR, 1957)

are newly flight mature and fly at most for 2 h. The model developed by JOHNSON *et al.* (1957) enables very accurate predictions to be made of the shape of the flight curve on a particular day. Therefore wind and humidity, which have previously been regarded as important factors in aphid flight, can be completely ignored when considering flight periodicity in the black bean aphid. In fact JOHNSON (1969) has produced convincing evidence that most aphid migration takes place in windy weather.

Many aphids show the same bimodal flight periodicity as the black bean aphid and perhaps the model is applicable to these. However, the lime aphid and also some other species have a unimodal flight periodicity and the peak in number flying occurs just after midday (Fig. 4–2b). The lime aphid has a rather high temperature threshold for flight. The lowest temperature at which flight has been observed in the laboratory for this species is 18°C, and 50% of the flights occurred at temperatures in excess of 26°C. Therefore, the unimodal flight periodicity in this case may just be a consequence of the very high temperatures necessary for flight, which are only likely to be experienced regularly in the early afternoon. The sycamore aphid has a bimodal flight periodicity (Fig. 4–2c) with peaks coming early in the morning and after sunset. This species has a very low temperature threshold for flight, 50% of the flights occurring at temperatures below 11°C and so the aphid rarely experiences temperatures too low for flight, especially during the day. In common with other tree dwelling aphids the sycamore aphid takes off less readily than the black bean aphid in wind velocities of 8 and even 4·8 km h⁻¹ (HAINE 1955). Most sycamore aphids fly in very calm conditions. The peak of flight

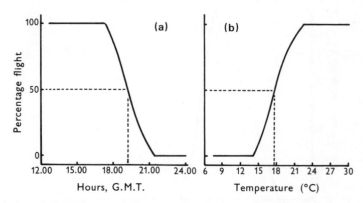

Fig. 4-4   Threshold of flight response to (**a**) light and (**b**) temperature in the black bean aphid obtained by plotting the percentage occasions that flight occurred against time and the maximum temperature. The light intensity at 19.15 G.M.T. is approximately 100 foot candles (1076 lm/m^2). (After TAYLOR, 1963)

activity in the morning and evening is associated with low wind speeds and the trough in the flight curve is associated with high wind speeds.

By flying when wind speed is very low the sycamore aphid, possibly in common with other tree dwelling aphids, can retain control of the result of its flight movements and remain within sight and reach of plants on which it can alight. Let us speculate why the sycamore aphid should find it advantageous to fly most readily when wind velocity is low. Natural woodlands are perennial and often cover a large area. Directed flight may therefore be the most effective means of dispersal from tree to tree. Many annual and short-lived plants provide only temporary habitats and tend to be irregularly distributed in time and space. Aphids dependent on turbulent convective processes of air movement for distribution can cover much greater areas, and therefore have a better chance of finding other annual and sporadically distributed hosts.

4.3   Energy for flight

Energy for flight is obtained by the respiration of fats and glycogen. Glycogen is used first and then fat becomes the principal fuel after the first hour of flight (Fig. 4-5). In respiring fats water is released which enables the aphid to maintain its water content even after prolonged flight in dry air. The fuel reserves of the black bean aphid are sufficient to support flight for up to 12 h (COCKBAIN, 1961a). Therefore aphids have adequate fuel for the 1-3 h migratory flight postulated by JOHNSON (1969).

Plate 3 *Tuberolachnus salignus* (Gmelin) giving birth.

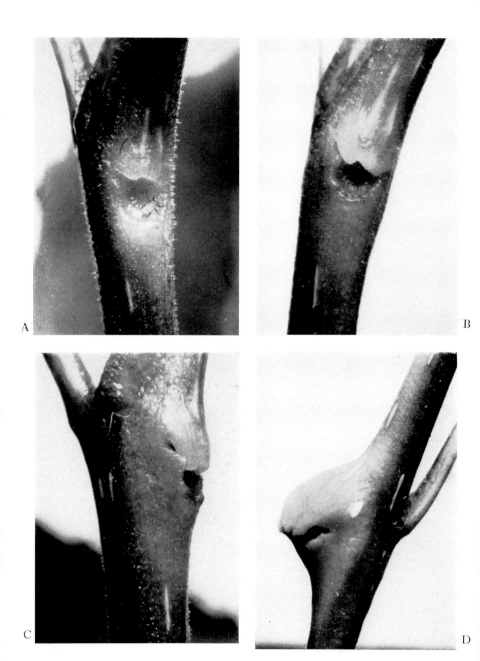

Plate 4 Development of the purse-like gall induced by *Pemphigus bursarius* L. A, Hollow develops on the petiole where the fundatrix probes. B & C, The edges of the hollow proliferate to form two lips. D, Fifteen days after a fundatrix begins probing the two lips have grown together to enclose the aphid. (Photographed by Dr. J. A. Dunn)

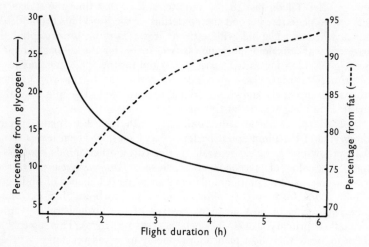

Fig. 4–5   The percentage of glycogen and fat present in alates of the black bean aphid after flying for from 1 to 6 hours

4.4   Location of host plants

Aphids do not fly to exhaustion before alighting and they take flight again if they settle on a non-host plant. This is essential if host plants are to be found, since some species of aphid alight indiscriminately on vegetation and they seldom find a host plant the first time they alight.

With one known exception, attempts to establish olfactory attraction of aphids to host plants have given negative results. Many aphids alight and probe in response to long wave length light reflected from the ground and vegetation (KENNEDY, 1961b). The predominantly long wave light emission from both leaves and soil contrasts sharply with the short wave light emission from clear or clouded skies. Aphids taking off for the first time show a strong attraction to short wave light which takes them up into the sky and away from plants. After a period of upward flight the attraction of short wave light wanes and aphids are then attracted to sources of long wave light. In the field aphids alight preferentially on leaves reflecting a greater proportion of the longer wave lengths of light, irrespective of plant species. Further the attraction of many species to yellow serves to draw aphids towards plants of the optimum physiological age: those that are actively growing or are senescent and therefore most nutritious. The much greater accumulation of aphids on a host plant in a mixture of vegetation is not due to their alighting selectively on the host. It is because aphids are likely to make a prolonged stay on a suitable host and to depart from a non-host (KENNEDY, 1959).

On landing, aphids invariably probe the surface. This initial probe is always brief, taking less than a minute, and in that time the stylets are unlikely to penetrate beyond the epidermis of the leaf. This probe determines whether an aphid will settle or leave. For the broom aphid the arrestant is the alkaloid sparteine (SMITH, 1966), for the cabbage aphid it is the glucoside sinigrin (WENSLER, 1962) and for the apple grass aphid on its primary host it is also a glucoside, phloridzin (KLINGAUF, 1971). These substances are of no known nutritional importance to the aphid but they enable it to recognize its host plant.

Not all aphids settle indiscriminately with regard to plant species. Emigrants of the plum aphid prefer to settle on grey-green leaves rather than yellow-green leaves, generally preferred by other aphids (MOERICKE, 1969). The colour of its secondary host, the reed, *Phragmites communis* Trin., is grey-green and so plum aphids have an increased chance of finding it. When sexual forms of the plum aphid return to plum they exercise a different means of host location. Sexuals of the bird cherry-oat aphid locate their primary host by its smell (PETTERSSON, 1970). This too enables them to locate their host plant more quickly in mixed vegetation.

Aphids that live on only one or two plants may find their host plants by responding to a specific colour or scent. Polyphagous aphids are less selective about alighting on a plant. However, by alighting on yellow leaves they select young growth or senescent leaves, in either case a highly nutritious food source.

4.5   Wing muscle autolysis

After reproduction has begun, in some species of aphid the indirect flight muscles begin to degenerate. This effectively prevents further flights and it has been suggested that the break-down products from the muscles are used in reproduction. However, it is unlikely that the degenerating wing muscles of the black bean aphid, which constitute 13% of the aphid's weight, can supply more than a small proportion of the material needed for the 27 developing embryos, which at birth are each 6% of their mother's weight. The breakdown of the flight muscles is very rapid in the vetch aphid, but in other species it can take 14 to 21 days. In the vetch aphid 50% of the volume of indirect flight muscle fibril is lost within two days of settling, which indicates a programmed destruction of the muscle by cytolysis. Some tree dwelling aphids do not lose their wing muscles and this is possibly related to their very long life as an adult during which they are likely to be exposed to a wide range of favourable and unfavourable conditions.

4.6   Fecundity of migrants

During their reproductive life alate aphids produce fewer and smaller offspring than apterous aphids of the same species. Alate aphids produce their offspring earlier in adult life than do apterae. To produce their offspring early in adult life ensures a more rapid succession of generations and an increase in numbers in the population. Therefore, although alatae are less fecund they are nevertheless able to contribute to a high rate of increase, and in some cases it is as high as that of the apterae.

The migratory flight affects subsequent reproduction in alates of the vetch aphid. If they fly they produce 16% fewer offspring than either alatae that do not fly, or apterae. Therefore, the act of flying influences the number of offspring produced. However, in the black bean aphid long migratory flights do not affect the reproductive potential of those aphids that fly (COCKBAIN, 1961b).

The reproductive physiology and behaviour of alatae is well adapted to exploit changes in the aphid's environment. Changes in the aphid's environment induce the appearance of alatae but whether they fly is determined by conditions experienced during their development and especially when they reach maturity. Should they fly then the act of flight has a marked effect on their subsequent behaviour and reproduction.

5 Aphid–Plant Interactions

5.1 Introduction

Aphids feed on the phloem sap of plants. The quality of the phloem sap changes with the progress of growth and maturation of the leaves and shoots, and the effect of these changes on the well-being of aphids has already been discussed. Aphids, however, can also interfere in quite dramatic ways with the growth processes of plants and it is this aspect of the aphid–plant interaction that we now consider.

5.2 Galling of plants

Some aphids induce the development of local structural abnormalities of their host plants. The form of these is characteristic for the species of aphid. Blister or pocket galls on the leaves of red currant, *Ribes rubrum* L., are caused by the currant aphid. Young leaves of apple infested with the rosy leaf-curling aphid twist and develop a red colouration. These malformations of the plant never completely enclose the aphid and are called pseudo-galls. In true galls the plant tissue grows around and completely encloses the aphid and its progeny. Later in the season the gall splits open and the aphids leave to find another host plant. Galls show an amazing variation in form and colour. Closely related species of aphid feeding on the leaves of the same host plant can induce the development of markedly different galls. The petiole of the leaves of poplar, *Populus nigra* L., may accommodate a purse-like gall induced by *Pemphigus bursarius* L., or a spiral gall induced by *P. spirothecae* Pass., or, the midrib of the leaf lamina may have a pouch-like gall caused by *P. filaginis* (B.d.f.) (Fig. 5–1).

The intricate behaviour of the three species of *Pemphigus* while inducing their galls is described by DUNN (1966). In all three species the fundatrix which initiates the gall seeks out a rapidly growing leaf, even though such leaves are wet with a sticky exudate which may trap the aphid. *P. bursarius* which induces a purse-like gall (Fig. 5–1a) inserts its stylets repeatedly over a very small area of the petiole of a leaf. The stylets remain inserted in any one position for approximately 10 min. Cell growth immediately around the punctures appears to be inhibited, whereas growth is stimulated in the tissue immediately surrounding this area. The area of arrested growth is engulfed as the petiole increases in diameter and more especially as the cells in the surrounding tissue proliferate (Plate 4). Once a hollow has developed the fundatrix moves around the periphery of the hollow and again stabs the tissue with its stylets. As a result the edges of the pit proliferate and form two lips which grow together and enclose the fundatrix in a hollow swelling within the petiole (Plate 4). The gall continues to increase in size for some time and its continued development depends on the presence of the aphid within the gall.

P. spirothecae initiates its spiral gall on the petiole of a leaf (Fig. 5–1b) by inserting its stylets at intervals, like *P. bursarius*, but over a wider area of a petiole and in a zigzag pattern. This results in the petiole bending towards the band of cells in which growth has been inhibited, as cell growth on the opposite side of the petiole is normal. As the inhibited zone of growth zigzags across the petiole the latter bends to form a loose spiral of 3 coils. The fundatrix then stimulates each coil to grow sideways into the adjacent coils. Longitudinal growth of the coils increases the size of the gall.

The wart-like galls induced on a leaf lamina by *P. filaginis* (Fig. 5–1c) are induced by the aphid inserting its stylets at intervals whilst slowly moving in a short, circular or rectangular course between the major veins on the abaxial surface of an expanding leaf. Cells along the track taken by the aphid stop growing and the encircled area expands to swell out in an adaxial direction. In this way a pocket is formed around the aphid. The aphid stabbing with its stylets induces the pocket to increase in size, and normal elongation of the leaf stretches out the ring of non-growing cells around the mouth of the pocket drawing the sides together and enclosing the aphid completely.

For the three species of *Pemphigus* galling poplar, therefore, the form of the gall can mainly be attributed to the behaviour of the aphids as they initiate their respective galls rather than to a difference in the cecidogenic substances injected by the aphids. The variety of galls is largely attributable to the variety of plant parts available to aphids and the behaviour of the aphids initiating the gall.

How do aphids induce the tissues of a plant to develop into these bizarre forms? As they feed the aphids secrete large volumes of saliva into their host plant. There are two components to the saliva: that which forms the salivary sheath (p. 3) and is deposited in the plant tissues around the aphid's stylets, and, a watery saliva that is secreted directly into the sieve tube on which the aphid is feeding. Something in the saliva induces the development of a gall. It is not known what. MILES (1968) has suggested that aphids produce their own specific chemical organizer of plant growth, or even unspecific plant hormones. It is also known that indole acetic acid (IAA), which is a plant growth hormone is present in the saliva of gall-forming aphids (SCHÄLLER, 1968). The chemical composition of the saliva of plant sucking bugs is, however, partly determined by the composition of their food, and therefore high levels of IAA in the aphid and its saliva may only reflect high levels of IAA in the aphid's food when feeding on galled plant tissue. Alternatively, substances in the aphid's saliva could inhibit the action of the plant's IAA oxidase and allow the

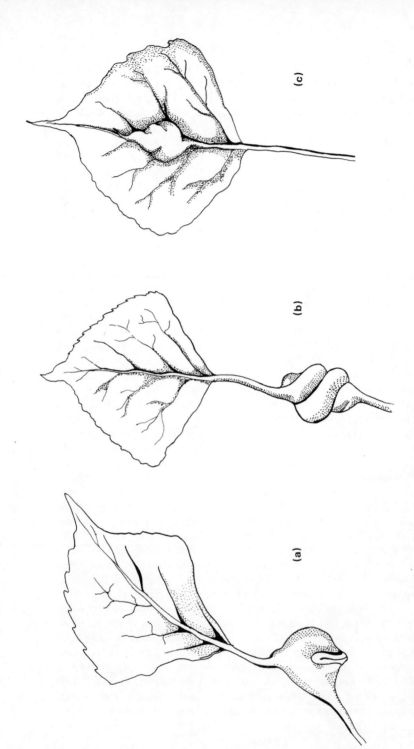

Fig. 5-1 The galls induced on the leaves of poplar, *Populus nigra* L., by (a) *Pemphigus bursarius* L. (b) *P. spirothecae* Pass. and (c) *P. filaginis* (B.d.f.)

level of IAA in the plant to rise. Now that many species of aphid can be reared on chemically defined diets, it must only be a matter of time before it is known what aphids put into plants in their saliva, and this will help to resolve just how aphids are capable of producing galls.

5.3   Advantages of galling

In galling a plant aphids obviously create a sheltered place in which to live and in true galls, an environment which is relatively free of insect parasites and predators; but it has also been shown that there is a local disturbance in the metabolism of the plant which results in an improvement in the aphid's food supply. KENNEDY (1958) everted blister galls of red currant and observed that aphids disturbed from their original feeding place show a marked preference for the part of the leaf that previously formed the interior of the gall, regardless of its exposure. Although this does not exclude the possibility that the gall also serves to protect the aphid it does suggest that physiological changes in the galled part of the leaf constitute the main advantage to aphids, confining their feeding and reproductive activities to the interior of galls. As gall forming aphids always induce the development of galls it is impossible to rear the aphid on an ungalled plant to test for the differences between aphids living in galls and aphids without galls. However, FORREST has shown that a non-galling aphid indigenous to the same host plant benefits by feeding on galled tissue produced by the gall forming species, and presumably the gall forming aphid benefits in the same way from the gall it produces. Galls provide improved nutrition for the aphid so that it can reach a greater size and produce more offspring. This improvement is temporary but does serve to extend the period favourable for the aphid's increase in numbers (FORREST, 1971).

5.4   Effects on plants other than galling

Apart from local changes caused by galling, extensive although often less conspicuous damage occurs in other parts of the plant. In addition to galled leaves, apple seedlings infested with the rosy leaf-curling aphid have thickened stems and abnormal root and shoot growth. If aphids are made radio-active by feeding them on radio-active plants, and these radio-active aphids are transferred to feed on normal plants, the new plants also become radio-active. This demonstrates very clearly that substances in aphid saliva can be translocated throughout the plant from the topmost leaves to the smallest roots. Therefore, it is not surprising if the growth of all parts of an aphid infested plant is abnormal. The effect of removing a ring of phloem from around the stem of the plant shows that the radio-active component of saliva is translocated in the phloem (Fig. 5–2; cf. RICHARDSON, 1968).

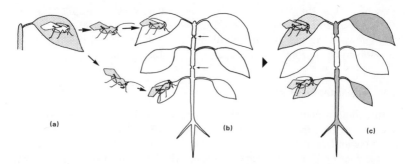

Fig. 5–2 Aphids made radio-active by feeding on radio-active plants
(**a**), when transferred to feed on normal plants (**b**) make these plants radio-
active (**c**). Removal of a ring of phloem tissue from around the stem of the
plant (←) prevents the translocation of radio-active material to other
parts of the plant

Many aphids do not induce the development of galls, although when
present in large numbers, aphids may even cause the death of part or the
whole of the plant. Death of the plant has been attributed to the energy and
nutrient drain inflicted by the aphids. However, there is now evidence to
support the idea that it is what the aphid puts into a plant rather than what it
takes out that contributes most to the death of part or the whole of the plant.
Both the black bean aphid and the cabbage aphid are capable of changing
the metabolism of their host plants in their favour. So although there are
none of the outward signs of galling there are restricted local changes in
the plant's metabolism similar to those that occur when the plant approaches
senescence. As with galling, these changes are to the aphid's advantage,
and the aphids which develop are as a result larger and more fecund.

Other plants show few or no obvious signs of aphid infestation. However,
careful study of these plants can reveal extensive changes. Trees like lime
and sycamore often support large numbers of aphids: a 14 m lime tree
can carry over a million aphids at one time. When the aphids are abundant
their excreta (honeydew) render the leaves of the trees and adjacent
vegetation very sticky. In a year the aphids on a 14 m lime can produce
31 dm³ of honeydew with a dry weight of 8·5 Kg. As this honeydew is
mainly sucrose, with amino-acids and minerals—the life sap of the plant
in fact—it is not surprising that aphids can affect the growth of a tree by
imposing a nutrient drain of such considerable magnitude.

Both lime and sycamore trees are infested by their own species of aphid,
Eucallipterus tiliae (L.) and *Drepanosiphum platanoides* (Schr.), respectively.
Aphid infestation in both lime and sycamore results in reduced growth,
and the more aphids infesting a tree the greater the reduction in growth
(Fig. 5–3). However, the energy drain imposed by aphids accounts for
only a third of the observed reduction in growth caused by aphids.

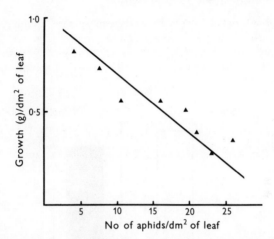

Fig. 5-3   The relationship between the growth of lime and the number of aphids infesting a tree

Although both aphid infested lime and sycamore grow less than uninfested trees the aphids affect them in different ways. In sycamore, aphid infestation results in a reduction in the size of the leaves, and also of the length of the twigs and the roots. The leaves also become a darker green than normal as they contain more chlorophyll. In lime, aphid infestation does not reduce the area of the leaves, nor shoot growth, but seriously inhibits root growth. In addition, the leaves contain less chlorophyll, yellow and fall earlier than the leaves of non-infested trees.

Associated with their greater chlorophyll content, the leaves of aphid infested sycamore trees produce considerably more carbohydrate by photosynthesis than do the leaves of uninfested trees. In this way sycamore is able to adapt in part to the energy drain inflicted by the aphids. Lime leaves, however, are unable to compensate in the same season for the energy drain imposed by the aphids, possibly because the aphids damage the leaves; the leaves yellow and fall earlier from infested trees. However, if uninfested in the following year the new crop of leaves contains more chlorophyll, and is a darker green and fixes considerably more energy than the leaves of previously uninfested trees (DIXON, 1971b and c).

The whole plant is affected by an aphid infestation even though parts of the plant may be distant from the site of infestation. This results from the translocation throughout the plant of substances injected into the plant by the aphid when it feeds. It is also possible that the drastic effect of aphids' saliva on the growth of lime and sycamore is aggravated by the effect of the large quantities of sap removed by the aphids.

5.5   How aphids affect plant growth

Infestation of radish plants by the peach-potato aphid results in a marked disturbance in the concentration of the plant's growth hormones. There is a pronounced increase in the level of the plant growth inhibitors and a decrease in the growth promoters (Fig. 5–4). This imbalance could

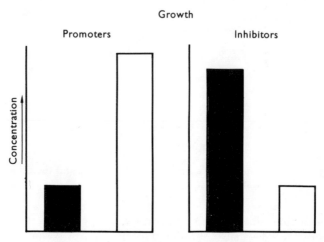

Fig. 5–4   The concentration of plant growth promoters and inhibitors in aphid infested (■) and aphid free (□) radish seedlings

result from the plant growth inhibitors introduced into the host plant via the saliva of the aphid. Possibly this effect is aggravated by the aphid removing from the plant large quantities of plant growth promoters and essential food materials.

By galling, aphids induce changes in plants which, although temporary, are to the aphids' advantage. However, the effect that certain tree dwelling aphids have on the growth of their host plants does not appear to be of any advantage to the aphid. In sycamore, severe infestation of a tree in spring results in poor growth of the leaves and shoots. This is not to the aphid's advantage as it thrives when the plant is actively growing.

The changes induced in a plant by aphids do not always improve conditions for the aphid, and one can still only guess some of the functions of the saliva that is injected into the phloem elements of plants. In piercing a phloem element and removing phloem sap an aphid causes a drop in turgor at the point of puncture. The normal response of a plant to such a wound is to seal off the phloem elements concerned and bleeding then stops. However, the flow of sap for the aphid is not interrupted in this way even when large numbers of aphids are all feeding together, which would be equivalent to inflicting a major wound. It is possible that one function of the watery component of the aphid's saliva is to prevent or reduce the intensity of the normal response of a plant to a wound. In doing this the aphid would be assured of a continuous flow of phloem sap, but indirectly it could adversely affect the growth processes of the plant. It is possible that the substances involved in the production of the salivary sheath also damage a plant, and could account for the localized galling and senescence observed in some aphid–plant interactions.

6 Aphids and Plant Viruses

6.1 Introduction

More plant viruses are transmitted by aphids than by any other group of animals (SWENSON, 1968) and it is likely that aphids do more damage by transmitting viruses than by removing the sap of plants. Viruses are carried by migrating aphids which may travel as far as 1300 km. Very large numbers of alatae develop on aphid infested crops: as many as 580 000 per year may disperse to other crop areas from 0·4 hectares (an acre) of sugar beet (KERSHAW, 1964), and as many as 1600 million from an acre of field beans (WAY & BANKS, 1967). Thus aphids with the viruses they transmit are a potential threat to agricultural crops.

6.2 Virus and its aphid vector

Viruses transmitted by aphids can be classified according to the way in which the aphid carries them: the stylet borne and circulative viruses (KENNEDY et al., 1962). A few viruses are intermediate in their characteristics.

Aphids of any instar can acquire stylet borne viruses very quickly. After probing the epidermal and subepidermal tissue of an infected plant for only a few seconds it is possible for an aphid to transmit the virus immediately to other plants (Fig. 6–1a). The more an aphid probes among the epidermal cells whilst wandering over an infected plant the more likely the aphid is to become infective. However, infectivity only lasts an hour or two. If an infective aphid moults it loses its infectivity and it is assumed that the virus remains attached to some part of the aphid's exoskeleton that is shed in the moult. Irradiating the tip of an infective aphid's stylets with ultraviolet light or dipping the tip of the stylets into formaldehyde renders the aphid non-infective, and so it is supposed that the virus is carried only by the aphid's stylets and is introduced mechanically into a plant when the aphid probes in its search for a suitable feeding site (p. 27). Each stylet borne virus can be transmitted by a number of different species of aphid and is often to be found in several related plant species. An aphid may carry several viruses at once, stylet borne and circulative.

For an aphid to become infective with a circulative virus the aphid must actually feed from, not merely probe, an infected plant. Its stylets must penetrate deep into the tissues of a plant and pierce a sieve tube within the

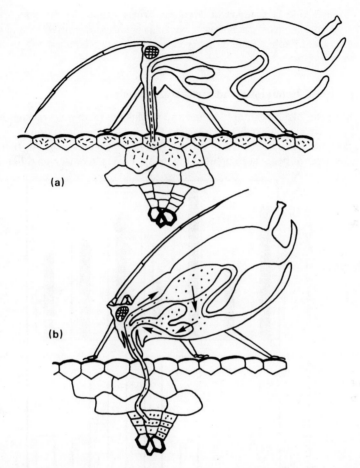

Fig. 6–1 Diagrams showing mechanisms of acquisition of viruses by aphids. (a) Stylet borne viruses acquired when aphids probe the epidermal tissues of infected plants. (b) Circulative viruses acquired when aphids feed on the phloem tissues of infected plants (′′′′, stylet borne virus and : : :, circulative virus. (Modified after WATSON, 1967)

phloem tissue (Fig. 6–1*b*). It therefore takes as long as several hours for an aphid to become infected. The virus pervades the body of the aphid and can be recovered from the aphid's body fluids. After acquiring circulative virus there is a latent period before the aphid becomes infective. During this period the virus multiplies in the aphid's tissues and enters the salivary glands, whence it can be injected into other plants along with the saliva. Once infective an aphid often remains so throughout its life and can transmit the circulative virus to a succession of plants. Moulting does not affect

an aphid's role as a carrier of circulative virus although it frees an aphid
of stylet borne virus. Each circulative virus is only transmitted by one or a
few species of aphid, unlike the stylet borne viruses which are not usually
vector-specific.

6.3   Aphid behaviour and virus transmission

Both the black bean aphids and the peach-potato aphids transmit beet
yellows virus to sugar beet. However, the incidence of infection in sugar
beet is more closely related to the number of peach-potato aphids than to
the number of black bean aphids (Fig. 6–2). This is because of a difference

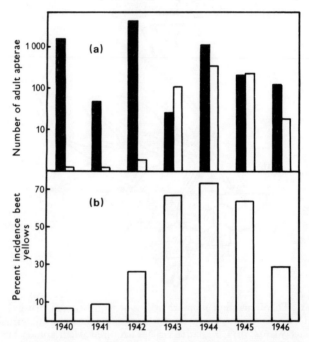

Fig. 6–2   (a) The number of black bean aphids (■) and peach-potato aphids
(□) infesting sugar beet, and (b) the incidence of beet yellows virus from
1940 to 1946. (After WATSON, 1967)

in behaviour between the two species. Peach-potato aphids are never very
numerous but they move about more over their host plants and disperse
quickly from plant to plant within a crop, thus spreading virus. In contrast
black bean aphids form dense aggregates and are sedentary so may only
be present on a few plants and remain on these.

Because of its restlessness the peach-potato aphid is also adept at spreading potato leaf roll virus (circulative) and potato virus Y (stylet borne)

within potato crops. The more numerous buckthorn aphid on potatoes rarely moves from plant to plant and is therefore less effective in transmitting virus within a crop (WATSON, 1967).

After moulting to become adult, aphids are unlikely to feed before they migrate. This should result in alatae, even from virus infected plants, being free of stylet borne viruses as these are lost when an aphid moults. However, inhibition of feeding is rarely complete. It was found that as many as 16% of alate peach-potato aphids reared on plants infected with beet mosaic virus were infective when they migrated, but alate migrants of the black bean aphids from the same plants were not infective (COCKBAIN & HEATHCOTE, 1965). The degree of overcrowding and the state of the host plant influence whether or not an aphid will migrate (p. 22). The same factors may also determine the probability of probing before an aphid takes flight. This type of flight behaviour affects the transmission of stylet borne viruses only.

As many aphids alight indiscriminately with regard to plant species, and then probe to determine whether or not the plant is a suitable host, they are well suited to spreading viruses among closely related species of hosts. Virus can also be transmitted by nymphs and apterae which walk from plant to plant when aphids are abundant and similar plants are growing close together. The agricultural practice of monoculture especially lends itself to the spread of virus within a susceptible crop.

As aphids appear to be ideally fitted for spreading virus it is surprising that virus disease in agricultural crops is not more widespread (KENNEDY, 1960). However, virus does not usually survive either in the seed of plants or the egg stage of aphids and so its incidence is reduced annually. A reservoir of virus overwinters in biennial and perennial plants. The number of immigrant alates actually carrying virus is low in most species and in some species the tendency to aggregate and to remain sedentary restricts the spread of infection. The short flight period and subsequent autolysis of the flight muscles in many migrating aphids (p. 28) restricts the spread of circulative viruses which need time to develop within the aphid before it becomes infective, by which time the aphid has settled on a host plant and is likely to remain there. This is partly offset by the migration of large numbers of alates and their ability to multiply extremely rapidly in optimum conditions, and in some species, to move actively from plant to plant.

6.4 Weather and virus transmission

Most aphids that transmit virus in temperate regions are host alternating species (p. 17) and the trees and shrubs on which these aphids normally overwinter do not usually harbour viruses of agricultural crops. However, outbreaks of virus disease in agricultural crops are associated with mild winters and springs. The reason for this is that not all the individuals of some host alternating species return to their primary host in the autumn; some remain on their herbaceous hosts all through the year, without entering an egg stage during the winter. In mild winters the survival rate of these free living forms is high, and in spring they multiply early and migrate. The herbaceous plants on which these aphids overwinter are often infected with virus. Thus in years when alate aphids arrive very early on the crops they have overwintered on herbaceous hosts and are likely to be infected with virus, but the aphids which arrive later come from their woody hosts and are usually virus free.

For example, the incidence of beet yellows virus in sugar beet is associated with the number of days in January, February and March when the temperature falls below 0°C, and the mean weekly temperature in April. Very accurate predictions of the incidence of beet yellows virus can be made for a particular year (Fig. 6–3). The temperature in January,

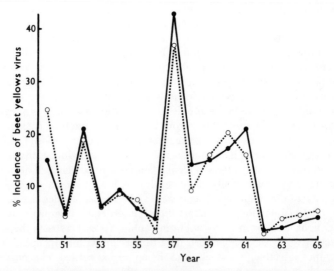

Fig. 6–3   The observed extent of beet yellows virus infection of sugar beet in August between 1950 and 1965 (●—●) compared with that predicted from previous winter and spring temperatures (o—o). (After WATSON, 1966)

February and March determines how many aphids survive the winter, and that in April influences the speed with which surviving aphids can multiply and migrate. The incidence of carrot motley dwarf virus is also influenced by the mildness of the winter and its effect on the survival of the anholocyclic clones (p. 9) of the carrot aphid. Mild winters result in outbreaks of this virus disease and consequently a severe reduction in carrot crops. The effect of weather on aphids also determines where certain crops are grown. Cooler conditions in Scotland do not favour rapid increase in aphids or the dispersal of aphids from plant to plant. As a consequence aphid borne viruses spread very slowly. Therefore, areas like Scotland are used to produce virus free plants, such as seed potatoes. If virus free plants are to be produced in areas where climatic conditions are more favourable for aphids they must be prevented from infesting the crops. For example, in the Netherlands virus free seed potatoes are produced by lifting the potatoes as soon as the aphids that transmit potato viruses begin to migrate.

6.5   Advantages of virus transmission to aphids

The amino-nitrogen content of the phloem sap of mature plants or parts of plants is generally low. This imposes severe restrictions on the rate of growth and fecundity of aphids. Virus infection of plants results in increased concentration of free amino-acids. As a result of this improvement in the quality of their food aphids develop faster and give birth to more young on virus infected plants. Aphids do not appear to be adversely affected by feeding on virus infected plants or by carrying virus. Therefore, it would appear as KENNEDY (1951) suggested that at least in the short term aphids have much to gain transmitting plant viruses. The benefit is mutual since the aphid is the means of dispersal of the virus, and the virus improves the quality of the plant as food for the aphid.

7 Predators and Parasites of Aphids

7.1 Introduction

Of the many predators of aphids most are insects; they include the larvae and adults of lacewings and most ladybird beetles, the larvae of some hoverflies, and cecidomyid larvae. Certain birds also eat aphids, especially when these are abundant or when other food is scarce. Aphids are also attacked by fungus infections and hymenopterous parasites. The hymenopterous parasites insert their eggs into the body of aphids and the parasite larva develops within the host, finally killing it. The parasite larva sticks the skin of the dead aphid to the surface of a leaf and then spins a cocoon within it. This cocoon, together with the skin of a dead aphid, is called a 'mummy'. Only one parasite develops to maturity in each parasitized aphid.

The complex of insect parasites and predators associated with the

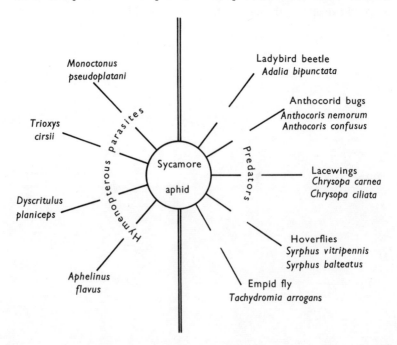

Fig. 7-1 The predators and parasites which attack the sycamore aphid

sycamore aphid is given in Figure 7–1. Insect predators of aphids will feed on a number of different species of aphid but the insect parasites are more specific and are often restricted to a single species of aphid. The hyperparasites which attack the parasites, however, usually parasitize a number of different species of primary parasites.

7.2 How insect predators and parasites find aphids

Insect predators and parasites of aphids are associated more with a particular habitat than with a particular aphid, and within a habitat they prefer to search for aphids in vegetation of a particular height. For example the adult of the hoverfly, *Syrphus lasiophthalmus* (Zett.), occurs in spring when aphid infestations are mainly to be found on trees and shrubs and prefers to oviposit at a height of 180 cm. *S. luniger* Mg. is common as an adult throughout the year and does not oviposit at any particular height, exploiting aphids wherever it can find them. In summer aphid populations on trees are failing or are less available, whereas many aphids are developing on herbaceous plants and hoverflies that occur only then, e.g. *Leucozona lucorum* (L.), prefer to oviposit at a height of 30 cm, exploiting the aphids on herbaceous plants (Fig. 7–2). Predators and parasites seem to search the vegetation at random in the preferred part of the habitat. They only appreciate the presence of an aphid when they are very close to their aphid prey.

Adult ladybird beetles are very inefficient at capturing aphids and can-

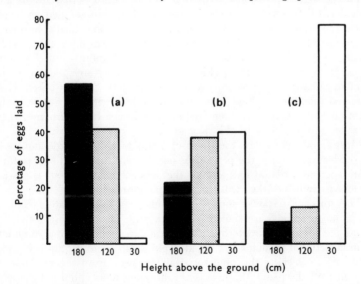

Fig. 7–2 The preferred height for oviposition in (**a**) *Syrphus lasiophthalmus*, (**b**) *Syrphus luniger* and (**c**) *Leucozona lucorum*. (Data from CHANDLER, 1968)

not appreciate the presence of an aphid until they touch it. If few aphids are present the beetles cannot eat sufficient food to mature their eggs and they quickly leave and seek aphids elsewhere. Ladybird beetles oviposit close to aphids because they are less active when well fed, and consequently remain in the vicinity of their food supply. They also lay more eggs when food is abundant. In this way eggs are more likely to be deposited close to large numbers of aphids, and the very small larvae that hatch from the eggs are more likely to be able to obtain sufficient aphids as food on which to survive (DIXON, 1959).

Adult aphidophagous hoverflies eat pollen to mature their eggs. They hover preferentially close to green stems. The smell of live or dead aphids, aphid skins and even honeydew, induces the hoverfly to alight and oviposit. Some species are even able to exploit concealed aphid colonies such as those in pseudo-galls and galls, or even on the roots, as well as more exposed colonies on the undersurface of leaves and stems (DIXON, T. J., 1959).

Thus ladybird beetles and hoverflies both lay their eggs close to concentrations of aphids although they achieve this in different ways. In both cases, however, the larvae do not have to go very far to reach the vicinity of their first prey, and the number of prey available in the area is often such that the predator can reach maturity.

7.3 Aphids' escape responses and defence mechanisms

The general impression conveyed by the literature is that aphids and related small insects are helpless, sedentary and thin skinned creatures that invite the attention of any predator that comes along. That aphids have defensive mechanisms has nevertheless been known since 1891, when BÜSGEN reported that several different species of aphid exuded an oily liquid from their siphunculi (Fig. 1–4), which they smeared on to the head of an attacking predator, and then escaped. Other workers have observed that some aphids, when disturbed, leap with alacrity from the plant on which they have been feeding, and escape from certain of their parasites and predators by this means.

Most aphids feeding on the stems of plants face downwards and those on the leaves, towards the petiole. As insect predators tend to move up a stem, and outwards on to a leaf via the petiole, this behaviour is advantageous to aphids as they will see predators approaching.

The sorts of response shown by aphids to insect predators, and of predators to aphids, are shown in Figure 7–3. The relative sizes of the aphid and its predator determine to a great extent the outcome of an encounter between predator and aphid. Aphids can avoid capture by insect predators by kicking the predator, walking away from the predator or dropping off the plant. While it is advantageous to an aphid to be able to repel a predator and continue feeding, it can only do this if the predator is small relative to the aphid. If the predator is not deterred by an aphid

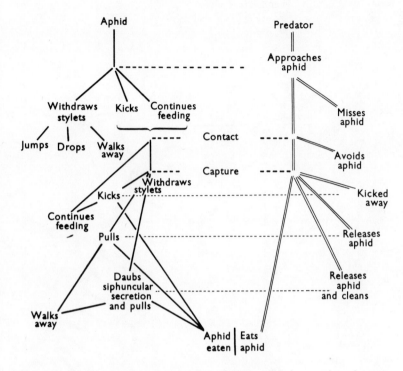

Fig. 7-3 Diagrammatic representation of the possible interactions between an aphid and a predator during an encounter

kicking, and continues to approach, the aphid's most effective means of escape is to withdraw its stylets hastily and walk away. Large predators walk faster than adult aphids and then it is expedient for the aphid to drop or jump off the plant, although the aphid may then die of starvation before it is able to find a host plant again.

If a predator seizes an appendage of an aphid the latter may attempt to kick the predator away, or pull the appendage free. If the predator is small relative to the aphid then the predator can often be kicked off the plant. In some aphids, if pulling fails, the aphid sheds the leg held by the predator or daubs secretion from the flexible siphunculi on to the predator's head (Fig. 7-4). The predator then becomes occupied with removing the siphuncular secretion and pulling is then frequently effective in enabling the aphid to escape. The siphuncular secretion is only effectively used as a defensive mechanism when the predator and aphid are similar in size (DIXON, 1958). Hymenopterous parasites are also frequently seen stuck to the siphuncular secretion and it may be a more effective means of protection against parasites. When a sycamore aphid is captured by a predator,

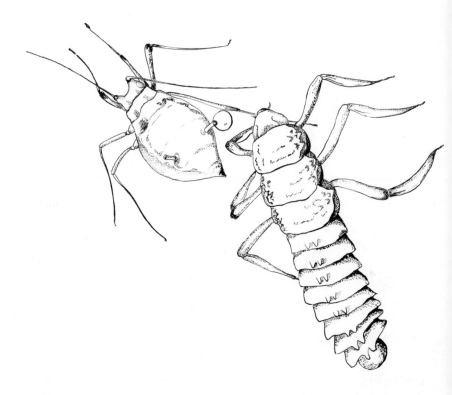

Fig. 7–4 An aphid attempting to escape from a larva of a ladybird beetle which has seized a hind leg of the aphid. The aphid attempts to pull its leg free, at the same time daubing a droplet of siphuncular secretion on to the head of the predator

the aphid often produces the secretion from its siphunculi and other aphids close by withdraw their stylets and walk away, in response to the smell of the siphuncular secretion and the view of the struggling aphid. In the peach-potato aphid the smell of the siphuncular secretion alone will make other aphids walk away or drop off the plant. The alarm substance that initiates this response is thought to be trans-β-farnesene (BOWERS *et al.*, 1972). In producing siphuncular secretion aphids may secure their own escape and also effectively warn others of the presence of a predator.

Aphids have a greater chance of avoiding capture when a predator approaches from the front, probably because they see it coming. An aphid is more likely to be caught when a predator approaches from the rear (Fig. 7–5). Therefore, the position of the siphunculi on the posterior part

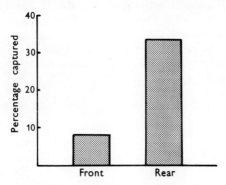

Fig. 7–5 The percentage of aphids captured when approached from the front or rear by a predator

of the aphid's abdomen is of advantage to the aphid in its attempt to escape once caught.

Aphids living on exposed parts of plants have long flexible siphunculi whereas those protected in galls, or on roots, or which tend to be hirsute or covered with flocculent wax, have short or rudimentary siphunculi which are not as valuable as protective devices. Thus the degree of development of the siphunculi varies, and those with short or rudimentary siphunculi usually have other means of protection.

Aphids exhibit two main trends in the evolution of their behaviour in relation to enemies. Many species depend chiefly upon their activity to avoid an approaching enemy. These aphids are also characterized by cryptic colouration and often form diffuse colonies. In contrast, inactive species do not avoid approaching enemies and are myrmecophilous and often conspicuous as the result of both their colouration and tendency to form large compact colonies. Aphids which allow ants to attend them are protected against predators and parasites, as ants will attack and remove these enemies from aphid colonies. However, both adults and larvae of the ladybird beetle, *Coccinella divaricata* Olivier, habitually feed on aphid colonies attended by *Formica rufa* L. the large and voracious wood ant. Because of their special behaviour and the production of secretions attractive to ants, a few highly specialized parasites and predators, like *C. divaricata*, can avoid being attacked by ants. Some inactive species of aphids may be unpalatable. Predators attacking unpalatable aphids reject them as soon as they pierce the body wall of the aphid. In subsequent attacks on this aphid, the body wall is not pierced and the aphid is rejected, as soon as it is touched by the predator. The predator appears to learn as a result of its previous experience. Not only are some aphids distasteful, but some are even poisonous. Some ladybird beetle larvae will attack the vetch aphid and begin to feed on it. However, after 2 minutes the larvae release their prey and regurgitate their gut contents. Many of these larvae subsequently die, a few days later, even although offered other suitable species of aphid as food.

Aphids defend themselves against their enemies in a variety of ways, and the effectiveness of their defensive mechanism can only be judged in relation to the mode of life of the aphid.

8 Regulation of the Numbers of Aphids

8.1 Introduction

Aphids can multiply so rapidly that they soon reach pest proportions. The very short period of approximately 14 days required for an aphid to reach maturity, combined with a high fecundity (approximately 30 nymphs born to each aphid) indicates a very rapid rate of increase. Some aphids can achieve as many as 9 generations during the favourable period of a year, and so a single aphid could give rise to over 600×10^9 aphids. Each aphid weighs approximately 1 mg and therefore in a season one aphid could give rise to over 600 000 kg of progeny, equivalent to the weight of 10 000 men. This can only be achieved if weather is favourable; and parasites, pathogens and predators are absent; and space and food are unlimited. Each of these factors can inhibit population growth, and increase in numbers of aphids is normally checked under natural conditions.

8.2 Weather

(*a*) *Temperature* The rate of development of aphids, and their fecundity, are both greatly affected by changes in temperature. Low temperatures slow down development and reduce fecundity. For each species of aphid there is an optimum temperature for development and reproduction, and above this temperature the rate of development and fecundity are again reduced (Fig. 8–1). Some species of aphid, such as the sycamore aphid, are able to acclimatize to high and low temperatures. At low temperatures the sycamore aphid's basal metabolism increases and enables the aphid

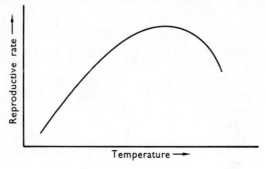

Fig. 8-1 The effect of temperature on the reproductive rate of aphids

to remain active even at quite low temperatures. At high temperatures the basic metabolism is lowered and as a result the aphid does not become over-active. However, this response is not shown by all aphids. Aphids that overwinter in an active stage rather than as an egg are drastically affected by cold in winter. Large quantities of lipoidal materials in the tissues of these aphids afford them some protection from freezing but do not protect them in very cold conditions. Outbreaks of the spruce aphid, which in Western Europe overwinters in an active form, are associated with mild winters which result in high survival of the aphid during the winter months. (*b*) *Rainfall* Aphids on the underside of leaves are to a large extent protected from the effect of rain. However, aphids feeding on buds or stems of plants are exposed and can be knocked off the plant in a heavy shower. The number of pea aphids is greatly reduced by a heavy shower (Fig. 8–2), because they tend to feed on the stems.

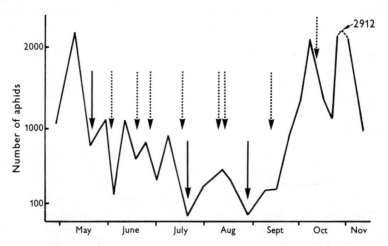

Fig. 8–2   The effect of heavy rain (↓) on the numbers of pea aphids on lucerne (↓ =crop cut). (After DUNN and WRIGHT, 1955)

(*c*) *Wind* It needs a very strong wind to dislodge an aphid on an isolated leaf. However, wind disturbs the foliage of plants and often leads to the leaves brushing together and dislodging many aphids. In high winds sycamore aphids move to the bases of the leaves where the veins protrude most and afford the aphid some protection, but strong winds can nevertheless result in many aphids being brushed off (DIXON & MCKAY, 1970).

Because of their cold-bloodedness aphids are greatly affected by changes of weather. Outbreaks of aphids can often be attributed to weather factors (BEJER-PETERSEN, 1962); i.e., weather has a density disturbing effect on aphid populations.

8.3 Aphid enemies

Attempts have been made to reduce the damage to agricultural crops by utilizing the enemies of aphids. Parasites (p. 44) have been used against the pea aphid and walnut aphid (HAGEN *et al.*, 1971; MESSENGER & VAN DEN BOSCH, 1971), and against the cotton aphid and peach-potato aphid on glasshouse crops (HUSSEY & BRAVENBOER, 1971). Ladybird beetles, the larvae and adults of which eat aphids, have been used against the alfalfa aphid (HAGEN *et al.*, 1971). Parasitic fungi have also been used to control aphids under glasshouse conditions. The use of biological control has been successful in reducing the incidence of the aphid pests in these cases. However, under natural conditions although they kill many aphids there is no evidence to indicate that enemies can regulate the numbers of aphids. In natural populations the percentage mortality due to predators and parasites in general is lower when the aphid population is large than when it is small. Thus they do not curb the increase of dense populations and they hinder the increase of sparse ones.

8.4 Intra-specific mechanisms

In highly polymorphic species of aphid a greater proportion of the aphids tends to develop into alatae as the aphids become more crowded (p. 14, Fig. 2–8). Crowding also results in the development of small adults with a low fecundity. Hence with increasing numbers proportionately more aphids fly away and those that are left each give birth to fewer offspring. A balance is reached when recruitment to the population due to births compensates for the loss due to migration and death (Fig. 8–3a). However, under field conditions aphid populations rarely stabilize but tend to decline rapidly after reaching high numbers (Fig. 8–3b). The population may again increase in numbers later in the year, but increase is then limited by low temperature and short day conditions which induce the development of sexual forms and the laying of overwintering eggs (p. 10). The sudden decline in number of some aphids is possibly due to the activity of the aphid's enemies, or a deterioration in the condition of their host plant. Insect enemies do accumulate on plants bearing large numbers of aphids, and high numbers of aphids can damage or kill plants. However, the typically dramatic decline in the number of aphids is often largely due to the departure of emigrant aphids, which have been produced in response to crowding and, or, changes in nutrition. After the number of aphids has suddenly dropped the aphids' enemies continue to remain abundant and eat and kill most of the remaining aphids, and often prevent them from multiplying significantly again that year.

It is possible that an overproduction of migrant forms, induced by crowding, is common, especially in the highly polymorphic species where the virginoparae can be winged or apterous. The offspring of winged aphids

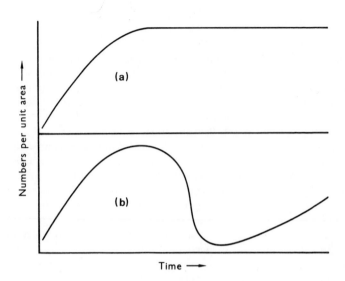

Fig. 8-3   Relationship between time and the number of aphids when (a) the population remains stable at a level determined by the carrying capacity of the environment, and (b) when the population reaches a high level of abundance and then declines and again increases later in the year

that colonize plants rarely develop into alatae. Inhibition of the production of alatae may extend over several generations. Therefore, after the initial colonization of a plant there is in highly polymorphic species a period when no alatae develop and the population grows. When the inhibition of the development of alatae has abated population density is likely to be high and many of the offspring then develop into alatae. There is an overproduction of alatae; more alates are produced than is necessary to stabilize the numbers of the aphid and there is a decline in the numbers of the aphid. This colonization and rapid exploitation of a plant, and then equally rapid exodus, possibly enables some species of aphid to avoid annihilation by their enemies or overexploitation of their host plants. The monophagous lime and sycamore aphids, whose parthenogenetic forms are all alate, also show a markedly overcompensated response to high numbers. In the lime aphid it mainly takes the form of emigration and in the sycamore aphid a reduction in fecundity.

Although parasites and predators often inflict heavy mortality on aphids it has only been claimed that enemies control aphid numbers in a few experimental cases, notably where they have been used in the biological control of certain pest species of aphid. It would appear that most of the aphids studied regulate their own numbers by dispersal and reduction in fecundity when density is high.

Scientific names of aphids mentioned in the text

Alfalfa aphid—*Therioaphis ononidis* (Kalt.)
Apple grass aphid—*Rhopalosiphum insertum* (Walk.)
Bird cherry-oat aphid—*Rhopalosiphum padi* L.
Black bean aphid—*Aphis fabae* Scop.
Broom aphid—*Acyrthosiphon spartii* (Koch)
Buckthorn aphid—*Aphis nasturtii* Kalt.
Cabbage aphid—*Brevicoryne brassicae* (L.)
Carrot aphid—*Cavariella aegopodii* (Scop.)
Cotton aphid—*Aphis gossypii* Glover
Currant aphid—*Cryptomyzus ribis* (L.)
Lime aphid—*Eucallipterus tiliae* (L.)
Pea aphid—*Acyrthosiphon pisum* (Harris)
Peach-potato aphid—*Myzus persicae* (Sulz.)
Plum aphid—*Hyalopterus pruni* (Geoff.)
Rosy leaf-curling aphid—*Dysaphis devecta* (Wlk.)
Strawberry root aphid—*Aphis forbesi* Weed
Sycamore aphid—*Drepanosiphum platanoides* (Schr.)
Vetch aphid—*Megoura viciae* Buckt.
Walnut aphid—*Chromaphis juglandicola* (Kalt.)
Willow aphid—*Tuberolachnus salignus* (Gmelin)

References

BEJER-PETERSEN, B. (1962). *Oikos*, **13**, 155–68.

BLACKMAN, R. L. (1971). *Bull. ent. Res.*, **60**, 533–46.

BOWERS, W. S., NAULT, L. R., WEBB, R. E., and DUTKY, S. R. (1972). *Science, N.Y.*, **177**, 1121–2.

BÜSGEN, M. (1891). *Jena. Z. Naturw.*, **25**, 339–428.

COCKBAIN, A. J. (1961a). *J. exp. Biol.*, **38**, 163–74.

COCKBAIN, A. J. (1961b). *J. exp. Biol.*, **38**, 181–7.

COCKBAIN, A. J., and HEATHCOTE, G. D. (1965). *Proc. Int. Cong. Entomol. 12th London 1964*, 521–3.

CHANDLER, A. E. F. (1968). *Entomophaga*, **13**, 187–95.

DAVIDSON, J. (1927). *Sci. Progr., Lond.*, **22**, 57–69.

DEHN, M. VON (1969). *Z. vergl. Physiol.*, **63**, 392–4.

DIXON, A. F. G. (1958). *Trans. R. ent. Soc. Lond.*, **110**, 319–34.

DIXON, A. F. G. (1959). *J. Anim. Ecol.*, **28**, 259–81.

DIXON, A. F. G. (1969). *J. Anim. Ecol.*, **38**, 585–606.

DIXON, A. F. G. (1971a). *Ann. appl. Biol.*, **68**, 135–47.

DIXON, A. F. G. (1971b). *J. appl. Ecol.*, **8**, 165–79.

DIXON, A. F. G. (1971c). *J. appl. Ecol.*, **8**, 393–9.

DIXON, A. F. G. (1972). *Entomologia exp. appl.*, **15**, 335–40.

DIXON, A. F. G., BURNS, M. D., and WANGBOONKONG, S. (1968). *Nature, Lond.*, **220**, 1337–8.

DIXON, A. F. G., and GLEN, D. M. (1971). *Ann. appl. Biol.*, **68**, 11–21.

DIXON, A. F. G., and MCKAY, S. (1970). *J. Anim. Ecol.*, **39**, 439–54.

DIXON, T. J. (1959). *Trans. R. ent. Soc. Lond.*, **111**, 57–80.

DUNN, J. A. (1966). *Marcellia*, **30**, 155–67.

DUNN, J. A. and WRIGHT, D. W. (1955). *Bull. ent. Res.*, **46**, 369–87.

FORREST, J. M. S. (1970). *J. Insect Physiol.*, **16**, 2281–92.

FORREST, J. M. S. (1971). *Entomologia exp. appl.*, **14**, 477–83.

HAGEN, K. S., VAN DEN BOSCH, R., and DAHLSTEN, F. L. (1971). Pp. 253–93 in *Biological Control*, edited by HUFFAKER, C. B., Plenum Press, New York and London.

HAINE, E. (1955). *Nature, Lond.*, **175**, 474.

HEIE, O. E. (1967). *Spolia zool. Mus. haun.*, **26**, 1–273.

HUSSEY, N. W., and BRAVENBOER, L. (1971). Pp. 195–216 in *Biological Control*, edited by HUFFAKER, C. B., Plenum Press, New York and London.

JOHNSON, C. G. (1969). *Migration and Dispersal of Insects by Flight*. Methuen, London.

JOHNSON, C. G., HAINE, E., COCKBAIN, A. J., and TAYLOR, L. R. (1957). *Ann. appl. Biol.*, **45**, 702–8.

JOHNSON, C. G., TAYLOR, L. R., and HAINE, E. (1957). *Ann. appl. Biol.*, **45**, 682–701.

KENNEDY, J. S. (1951). *Nature, Lond.*, **168**, 890–5.

KENNEDY, J. S. (1958). *Entomologia exp. appl.*, **1**, 50–65.

KENNEDY, J. S. (1959). *Ann. appl. Biol.*, **47**, 410–23.

KENNEDY, J. S. (1960). *Rept. Commonwealth Entomol. Conf. 7th, London* 165–8.

KENNEDY, J. S. (1961*a*). *Nature, Lond.*, **189**, 785–94.

KENNEDY, J. S. (1961*b*). *Ann. appl. Biol.*, **49**, 1–21.

KENNEDY, J. S., DAY, M. F., and EASTOP, V. F. (1962). *A conspectus of aphids a Vectors of Plant Viruses.* Commonwealth Inst. Entomol., London, 114p

KERSHAW, W. J. S. (1964). *Pl. Path.*, **13**, 90.

KLINGAUF, F. (1971). *Z. angew. Ent.*, **68**, 41–55.

LAMB, K. P. (1959). *J. Insect Physiol.*, **3**, 1–13.

LEES, A. D. (1963). *J. Insect Physiol.*, **9**, 153–64.

LEES, A. D. (1964). *J. exp. Biol.*, **41**, 119–33.

LEES, A. D. (1966). *Adv. Insect Physiol.*, **3**, 207–77.

LEES, A. D. (1967). *J. Insect Physiol.*, **13**, 289–318.

LEWIS, T., and TAYLOR, L. R. (1965). *Trans. R. ent. Soc. Lond.*, **116**, 393–479

MARCOVITCH, S. (1924). *J. agric. Res.*, **27**, 513–22.

MESSENGER, P. S., and VAN DEN BOSCH, R. (1971). Pp. 68–92 in *Biologica Control*, edited by HUFFAKER, C. B., Plenum Press, New York and London

MILES, P. W. (1968). *A. Rev. Phytopath.*, **6**, 137–66.

MITTLER, T. E. (1958). *J. exp. Biol.*, **35**, 626–38.

MOERICKE, V. (1955). *Z. angew. Ent.*, **37**, 29–91.

MOERICKE, V. (1969). *Entomologia exp. appl.*, **12**, 524–34.

MORDWILKO, A. K. (1928). *Ann. Mag. nat. Hist. (Ser.* 10), **2**, 570–82.

PETTERSSON, J. (1970). *LantbrHögsk. Annlr*, **35**, 381–9.

POLLARD, D. G. (1969). *Proc. R. ent. Soc. Lond.* (A), **44**, 173–85.

RICHARDSON, M. (1968). *Translocation in Plants*, Edward Arnold, London.

SCHÄLLER, G. (1968). *Zool. Jb. Physiol.*, **74**, 54–87.

SHAW, M. J. P. (1970). *Ann. appl. Biol.*, **65**, 197–203.

SMITH, B. D. (1966). *Nature, Lond.*, **212**, 213.

SUTHERLAND, O. R. W. (1969). *J. Insect Physiol.*, **15**, 2179–201.

SWENSON, K. G. (1968). *Ann. Rev. Phytopath.*, **6**, 351–74.

TAYLOR, L. R. (1963). *J. Anim. Ecol.*, **32**, 99–117.

TOBA, H. H., PASCHKE, J. D., and FRIEDMAN, S. (1967). *J. Insect Physiol.* **13**, 381–96.

TÓTH, L. (1940). *Ann. Mus. Hist.-nat. Hungar. Pars Zool.*, **33**, 167–71.

WATSON, M. A. (1966). *Pl. Path.*, **15**, 145–9.

WATSON, M. A. (1967). *Outl. Agric.*, **5**, 155–66.

WAY, M. J., and BANKS, C. J. (1967). *Ann. appl. Biol.*, **59**, 189–205.

WENSLER, R. J. D. (1962). *Nature, Lond.*, **195**, 830.

WHITE, D. (1971). *J. Insect Physiol.*, **17**, 761–73.